SUSTAINING A COMPETITIVE EDGE IN INNOVATION THROUGH A WORLD-CLASS FEDERAL SCIENCE AND TECHNOLOGY WORKFORCE

PRODUCT OF THE
Fast-Track Action Committee on the
Federal Science and Technology Workforce
OF THE NATIONAL SCIENCE AND TECHNOLOGY COUNCIL

July 2016

EXECUTIVE OFFICE OF THE PRESIDENT
NATIONAL SCIENCE AND TECHNOLOGY COUNCIL
WASHINGTON, D.C. 20502

July 27, 2016

Dear Colleagues:

I am pleased to transmit this report on *Leading a Competitive Edge in Innovation through a World-Class Federal Science and Technology Workforce*. The report is the result of a National Science and Technology Council Fast-Track Action Committee on the Federal Science and Technology Workforce (FSTW). Science and technology workforce opportunities and challenges are important across all S&T domains and agencies, and addressing them can have great positive impact on the Federal Government's work. Under its four-month charter, the FSTW—consisting of Federal S&T leaders and human capital professionals–developed strategic recommendations to help the government position itself as an attractive employer for next-generation S&T workers and to sustain existing Federal S&T talent. S&T stakeholders from across the Federal Government reviewed current agency efforts to support S&T workforce development including recruitment, retention, and succession planning for mission-critical S&T functions and shared innovative approaches that have been demonstrated to improve S&T workforce culture and environment, and opportunities for worker skills development when implemented.

The FSTW's recommendations are organized under four strategic priorities: focused effective leadership, S&T mission and workforce alignment, use of workplace authorities and flexibilities, and collaboration between human resources professionals and S&T managers. The recommendations provide valuable opportunities for the Federal agencies to maintain a world-class Federal S&T workforce; achieving these priorities, however, requires commitment to developing a personnel system and work environments in which the best and brightest are offered meaningful S&T careers in government service.

Your support of efforts in this direction will be much appreciated.

Sincerely,

John P. Holdren
Assistant to the President for Science and Technology
Director, Office of Science and Technology Policy

About the National Science and Technology Council

The National Science and Technology Council (NSTC) is the principal means by which the Executive Branch coordinates science and technology (S&T) policy across the diverse entities that make up the Federal research and development (R&D) enterprise. One of the NSTC's primary objectives is establishing clear national goals for Federal science and technology investments. The NSTC prepares R&D packages aimed at accomplishing multiple national goals. The NSTC's work is organized under five committees: Environment, Natural Resources, and Sustainability; Homeland and National Security; Science, Technology, Engineering, and Mathematics (STEM) Education; Science; and Technology. Each of these committees oversees subcommittees and working groups that are focused on different aspects of science and technology. More information is available at www.whitehouse.gov/ostp/nstc.

About the Office of Science and Technology Policy

The Office of Science and Technology Policy (OSTP) was established by the National Science and Technology Policy, Organization, and Priorities Act of 1976. OSTP's responsibilities include advising the President in policy formulation and budget development on questions in which S&T are important elements; articulating the President's S&T policy and programs; and fostering strong partnerships among Federal, state, and local governments, and the scientific communities in industry and academia. The Director of OSTP also serves as Assistant to the President for Science and Technology and manages the NSTC. More information is available at www.whitehouse.gov/ostp.

About the Fast-Track Action Committee on the Federal S&T Workforce

The Fast-Track Action Committee (FTAC) on the Federal S&T Workforce was chartered to coordinate strategic recommendations from the Federal agencies to enhance the Federal Government's efforts to recruit and maintain a world class S&T workforce by: enhancing the quality of the S&T workforce to execute S&T missions; reaching for and sustaining a diversity of S&T professionals that is representative of all segments of society; improving the flow of S&T experts into and out of Federal S&T mission agencies, so as to continually have a workforce that is matched to current and future technical and capability needs and that maintains awareness of a robust network of expertise outside of the Federal Government; and maintaining and building creative and dynamic work environments that offer good work/life balance and competitive compensation.

About this Document

This document was developed by the FTAC on the Federal S&T Workforce. The document was published by OSTP.

Acknowledgements

We would like to thank the committee that worked over several months to produce this report. Central to the committee's work was benchmarking and idea generation. A comprehensive list of those ideas can be found in the appendix of this report. Leaders across government may want to refer to the appendix for fresh and exciting ideas to achieve both the strategic goals of their own organizations as well as the goals of the Federal enterprise. Contributors that provided significant support to the activities of the Fast Track Action Committee and the development of this report include Asha Balakrishnan and Vanessa Peña from the IDA Science and Technology Policy Institute.

Copyright Information

This document is a work of the United States Government and is in the public domain (see 17 U.S.C. §105). Subject to the stipulations below, it may be distributed and copied with acknowledgement to OSTP.

Printed in the United States of America, 2016.

Report prepared by

NATIONAL SCIENCE AND TECHNOLOGY COUNCIL
FAST-TRACK ACTION COMMITTEE ON THE
FEDERAL S&T WORKFORCE

National Science and Technology Council

Chair
John P. Holdren
Assistant to the President for Science
and Technology and Director,
Office of Science and Technology Policy

Staff
Jayne B. Morrow (through June 7, 2015)
Executive Director
Afua A. N. Bruce (beginning June 8, 2015)
Executive Director

Fast-Track Action Committee on the Federal S&T Workforce

Co-Chair
Jeri Buchholz
Chief Human Capital Officer, Assistant
Administrator for Human Capital Management
National Aeronautics and Space Administration

Co-Chair
Jayne Morrow
Executive Director, National Science and
Technology Council
Office of Science and Technology Policy

Staff
William Ocampo
Program Analyst
Environmental Protection Agency

Members

Department of Agriculture
Diane McFadgen
Joon Park
Jennifer Riddle
Rosita Spears

Department of Commerce
Willie May
Kathryn Sullivan
Simon Frechette
Richard Cavanagh
Susanne Porch

Department of Defense

Alan Shaffer
Arati Prabhakar
Stephanie Barna
Kevin Kelley
Robin Staffin
Janie Mines
Col. Mary Lowe Mayhugh

Department of Energy
Patricia Dehmer
Cyndi Mays
Jeffrey Salmon
John Walsh

Department of Health and Human Services
Francis Collins
Nicole Lurie
Christine Major
Lawrence Tabak
Joe Martin
Sharon Ballard
George Korch

Department of Homeland Security
Reginald Brothers
Alitza Vega
Lisa Douglas Naughton

Department of the Interior
Suzette Kimball
Mary Pletcher
Kermit Howard

Department of Justice
Amy Hess
Pedro Espina
Jamie McDevitt

Department of Labor
David Michaels
Jonathon Bearr

Department of State
Judith Garber
Frances Colon
Raymond Limon
Olga Cabello Henry

Department of Transportation
Gregory Winfree
Ellen Partridge
Lydia Mercado

Department of Veterans Affairs
Ralph Paxton
Marisue Cody

Environmental Protection Agency
Angela Freeman
Robert Kavlock

Jim Johnson

National Aeronautics and Space Administration
Ellen Stofan
Gale Allen

National Science Foundation
France Córdova
Judith Sunley
Mayra Montrose

Nuclear Regulatory Commission
Jennifer Golder
Kristin Davis

Office of the Director of National Intelligence
David Honey
Deborah Kircher
Donna Call
Doris Johnson

Office of Management and Budget
Ali Zaidi
Dustin Brown

Office of Personnel Management
Katherine Archuleta
Mark Reinhold
Sydney Smith-Heimbrock
Jeanne Freidrich
Kimberly Holden

Office of Science and Technology Policy
Chris Fall

Table of Contents

Executive Summary 1

Background 3

Leadership to Sustain and Create an Inclusive and Innovative Workforce Enterprise 6

    Practice #1 – Enable Leadership for Inclusion 8

    Practice #2 – Promote Career Flexibility and Mobility 8

S&T Workforce Engagement in and Alignment to Mission 10

    Practice #1 – Align and Identify Critical Skills through Strategic Planning 11

    Practice #2 – Analyze Data and Assess Feedback 11

Effective Authorities to Enable a Flexible and Agile Workforce 14

    Practice #1 – Offer Competitive Pay 14

    Practice #2 – Convert Staff in Temporary Positions to Permanent Federal Service 15

Effective Relationships to Meet Strategic Workforce Needs 17

    Practice #1: Seek External Advice 17

Appendix: Full List of Recommendations 19

References 24

Abbreviations 25

Executive Summary

The Federal Government serves the Nation by promoting economic strength, preserving national security, and protecting public health, public safety, and the environment. Federal missions vital to the Nation's prosperity, health, and security are accomplished through the research and development and related activities conducted by the over 250,000 highly skilled scientists and engineers in the Federal workforce. As science advances and new technologies develop to meet national and global Grand Challenges[1], the science and technology (S&T) workforce needed to bring these technologies to bear must evolve. In building the future Federal S&T workforce, the Federal Government must consider trends in both the supply of and demand for these skilled professionals. First, the number of Federal employees retiring and the pool of retirement-eligible workers are increasing. Workers between the ages of 45 and 64 make up nearly 60 percent of the Federal S&T workforce. Second, the public perception of civil service has waned in recent years, and the Nation's pool of scientists and engineers is no longer as attracted to Federal missions as it was during the Space Race and Cold War eras. Third, wages for S&T workers in the Federal Government are frequently lower than in the private sector, especially for mid-career and executive leadership levels, and workplace flexibilities that are afforded to S&T workers in other sectors are often lacking. Finally, the workplace culture in other sectors may appear to some applicants to be more dynamic and vibrant.

The National Science and Technology Council (NSTC) established a Fast-Track Action Committee on the Federal S&T Workforce (FSTW) to examine these challenges and identify possible solutions. The recommendations provided by the FSTW are organized under four strategic priorities:

1. **Leadership for an Inclusive, Innovative Workforce Enterprise:** Leadership of agencies should promote a world-class science, technology, and engineering enterprise through thoughtful, proactive leadership and embrace strategies to create a culture of S&T inclusion and innovation within their workplace environments.

2. **S&T Workforce Engagement in and Alignment to Mission:** Agencies should cultivate strong connections between the Federal S&T workforce and their missions.

3. **Effective Authorities to Enable a Flexible and Agile Workforce:** Agencies should use successful human capital flexibilities, in combination with normal competitive hiring, to shape the Federal S&T workforce.

4. **Effective Relationships to Meet Strategic Workforce Needs:** Agencies should improve collaborations between S&T professionals and their human capital counterparts to strengthen human capital practices and enhance quality, diversity, and flow[2] within the Federal S&T workforce.

The implementation of the recommended practices within these priorities will require continued coordination and leadership across and within the Federal agencies.

Idea generation was central to the FSTW's work, and a comprehensive list of ideas can be found in the appendix of this report. The ideas are prioritized and coordinated with the FSTW-identified effective practices explored in the main report. Throughout the report, call-out boxes provide examples of current Federal efforts towards implementation of the FSTW's recommended practices. Leaders across

---

[1] Grand Challenges are ambitious but achievable goals that harness science, technology, and innovation to solve important national or global problems and that have the potential to capture the public's imagination. www.whitehouse.gov/administration/eop/ostp/grand-challenges.

[2] Flow refers to the movement of S&T workers across offices, agencies, departments and workforce sectors.

government may want to refer to the appendix for fresh and exciting ideas to help them achieve not only the strategic goals of their own organizations but also the goals of the Federal enterprise.

Background

Federal policies in support of the American innovation ecosystem encompass scientific research and development (R&D), a strong technical workforce, technology development, and commercialization and entrepreneurship across the enterprise. The Nation strives to increase innovative capacity and drive scientific and technical discoveries that lead to national economic growth and increased global competitiveness. S&T are critical to maintaining a healthy economy and preserving national security. The work of the Federal agencies provides the cornerstone for developing and implementing the policies and achieving these goals.

The work of the Federal agencies are diverse and scientifically and technologically rich. Scientists, technologists, and engineers in the Federal Government strive to catalyze breakthroughs in clean energy, health information technology, and precision medicine; stimulate innovative climate adaptation and mitigation strategies; spur productive entrepreneurship; and strengthen national security. Federal agencies support and conduct fundamental research and coordinate efforts to address the grand challenges of the twenty-first century. The NSTC recognizes that as science advances and new technologies develop to meet national and global Grand Challenges, the S&T workforce that will bring these technologies to bear must keep pace. Leadership within Federal agencies, academia, and industry must support a dynamic, diverse, world-class science, technology, engineering, and mathematics (STEM) workforce whose multi-disciplinary skills will provide the professional, technical, and policy expertise needed to solve the challenges of the Nation now and in the years to come.

The Federal agencies depend on a strong Federal S&T workforce to deliver their missions and advance innovation, but the Federal Government must consider trends in both the supply of and demand for these skilled professionals. Studies suggest an impending shortage of S&T-qualified professionals in specialty fields such as cybersecurity and intelligence and an increase in demand for general S&T skills from global industry.[3] A recent U.S. Department of Commerce report indicated that from 2000 to 2010, the number of STEM jobs grew 7.9 percent to 7.6 million, three times the growth rate of other fields.[4] The number of STEM jobs is expected to increase by 17 percent from 2008 to 2018, nearly double the rate of growth in other job categories.[5] This increase in demand is coupled with a limited supply of S&T professionals. The United States is expected to fall roughly 3 million people short of meeting the need for 22 million new workers with STEM-related postsecondary degrees by 2018.[6] The filling of STEM job vacancies is also complicated by the fact that STEM graduates are increasingly pursuing careers in non-STEM fields, such as business, finance, and law.[7]

Ongoing changes to the size and composition of the Federal S&T workforce may further negatively affect the Federal Government's ability to continue conducting and overseeing world-class R&D. First, both the

---

[3] National Academy of Engineering and National Research Council. *An Interim Report on Assuring DOD a Strong Science, Technology, Engineering, and Mathematics (STEM) Workforce* (Washington, D.C.: National Academies Press, 2012).

[4] D. Langdon, G. McKittrick, D. Beede, B. Khan, and M. Doms, *STEM: Good Jobs Now and for the Future*, ESA Issue Brief 03-11, July 2011.

[5] Paul Davidson, "Science, Tech Jobs Pay More, Lead in Growth," *USA Today*, July 13, 2011.

[6] A. P. Carnevale, N. Smith, and J. Strohl, *Help Wanted: Projection of Jobs and Education Requirements through 2018*, Washington, D.C.: Center on Education and the Workforce, Georgetown University, June 15, 2010.

[7] A. P. Carnevale, N. Smith, and M. Melton, *STEM: Science, Technology, Engineering, Mathematics* (Washington, D.C.: Center on Education and the Workforce, Georgetown University, 2011).

number of Federal employees retiring and the pool of retirement-eligible workers are increasing. Federal employees under 30 years of age account for only 6.9 percent of the Federal workforce and 7.5 percent of the Federal STEM workforce, while individuals between the ages of 45 and 64 make up nearly 60 percent of the Federal S&T workforce.[8] Although the Federal Government hired about 10,000 new STEM workers (about 6 percent of all new hires) in FY 2013, approximately 4,000 more STEM workers left the Federal Government than were hired.[9] A growing portion of Federal employees is nearing retirement-eligible age and is likely to leave the Federal workforce by 2020 through retirement or attrition.[10] The problem is especially acute for Federal scientists and engineers in the national security Federal S&T workforce, such as those in defense laboratories; in recognition, the recently released National Security S&T Strategy pays considerable attention to the challenges facing the national security component of the Federal S&T workforce and opportunities to revitalize it.[11] Filling these positions now and in the future is a challenge that must be addressed.

Second, public perception of the Federal personnel system as onerous can be a barrier to attracting and hiring top talent. Messaging about Federal S&T career opportunities is critical to attracting talent from the world's best science and engineering schools and high-tech companies and countering challenges in public perception.

Third, the Federal Government is a part of a fierce global competition for top S&T talent. For example, at best, 17 percent of current students say they will consider Federal service as part of their careers.[12] Some individuals who are initially attracted to Federal jobs leave them for positions in private industry, academia, and nonprofit organizations because those sectors offer more competitive pay, flexibility, or other benefits.[13] The Federal Government struggles to effectively engage the current Federal S&T workforce and apply its contributions to Federal S&T missions. Creating a work environment that enables innovative solutions to national challenges requires an awareness within the Federal leadership of the Federal S&T workforce and the varied roles its members play in achieving Federal missions.

Finally, the S&T enterprise relies on a workforce that is continuously changing in terms of demographics and geographical distribution. The Federal Government is uniquely positioned to embrace these changes. Federal agency assets and programs distributed across the Nation and around the globe are equipped with hiring and retention flexibilities that support work-life balance and foster flow and technical skill development where workforce capabilities are needed. New generations entering the workforce seek dynamic and fluid career opportunities and chances to apply foundational skills to challenging problems[14]. Those currently in the Federal workforce want to work for organizations that support creativity and innovation.[15] Realizing that a workplace environment that is inclusive and dynamic may address the loss of top S&T talent, Federal agencies are providing incentives and adapting workplace cultures to embrace

---

[8] Data from the Office of Personnel Management FedScope database on Federal human resources.

[9] Ibid.

[10] Department of Defense, *Department of Defense Science, Technology, Engineering and Mathematics (STEM) Education and Outreach Strategic Plan*, www.dod.mil/ddre/stem_education.html, 2009.

[11] National Science and Technology Council, *A 21st Century Science, Technology, and Innovation Strategy for America's National Security*, May 2016.

[12] Partnership and Universum USA, "Great Expectations! What Students Want in an Employer and How Federal Agencies Can Deliver It," http://ourpublicservice.org/OPS/publications/viewcontentdetails.php?id=131, 2010.

[13] A. P. Carnevale, *The Workplace Realities* (Alexandria, VA: AASA, 2008).

[14] OPM.gov "FEVS Millennials Report", available at http://www.fedview.opm.gov/2014FILES/FEVS_MillennialsReport.pdf.

[15] OPM.gov. "2014 Federal Employee Viewpoint Survey," available at http://www.fedview.opm.gov.

this incoming and transforming pool of talent. S&T workers may be attracted to or remain in Federal service if they are continuously given opportunities to enhance their skills and develop their careers in an environment that fosters and rewards retooling capabilities and building workforce capacity.

The NSTC established the FSTW to examine these issues and recommend solutions. The FSTW consisted of Federal S&T leaders and human capital professionals who provided strategic recommendations to help the government position itself as an attractive employer for next-generation S&T workers and to sustain existing Federal S&T talent. Achieving this goal requires commitment to developing a personnel system and a work environment in which the top talent are offered meaningful S&T careers in government service.

The recommendations provided by the FSTW are organized under four strategic priorities:

1. **Leadership for an Inclusive and Innovative Workforce Enterprise**: Leadership at agencies should promote a world-class science, technology, and engineering enterprise through thoughtful, proactive leadership and embrace strategies to create a culture of S&T inclusion and innovation within their workplace environments.

2. **S&T Workforce Engagement in and Alignment to Mission**: Agencies should cultivate strong connections between the Federal S&T workforce and their missions.

3. **Effective Authorities to Enable a Flexible and Agile Workforce:** Agencies should expand the use of successful human capital flexibilities to shape the Federal S&T workforce.

4. **Effective Relationships to Meet Strategic Workforce Needs:** Agencies should improve collaborations between S&T professionals and their human capital counterparts to strengthen human capital practices and enhance quality, diversity, and flow within the Federal S&T workforce.

These recommendations mirror, in many cases, the recommendations specific to the national security S&T workforce contained in the recently published National Security S&T Strategy.[16] Implementing the recommendations requires coordination and leadership across and within the Federal agencies. The FSTW identified effective practices that represent current Federal efforts towards implementation.

---

[16] National Science and Technology Council, *A 21st Century Science, Technology, and Innovation Strategy for America's National Security*, May 2016.

Leadership for an Inclusive and Innovative Workforce Enterprise

Leadership at agencies should promote a world-class science, technology, and engineering enterprise through thoughtful, proactive leadership that embraces strategies to create a culture of S&T inclusion[17] and innovation within their workplace environments.

Federal agency leaders at all levels (senior executives, S&T managers, etc.) are in positions to foster the S&T enterprise and champion the innovations and advancements of Federal scientists and engineers. They can use their unique platforms, both internal and external to their organizations, to positively influence the Federal S&T culture, promote inclusivity, and encourage workforce engagement to solve national challenges. The promotion of STEM education and careers as a means to create positive impact and to drive a culture of innovation is a national priority. The under-representation of racial and ethnic minorities in the S&T labor force is substantiated by the 2010 U.S. Census data, which documents the changing demographics of our Nation and the growth of minority populations. The literature suggests a strong association between innovation and high levels of racial, gender, and cognitive diversity.[18-19] To make the Federal Government the employer of choice for a world-class S&T workforce, Federal leaders must inspire top talent across this changing demography to enter S&T careers and Federal service.

Creating an innovative and inclusive workplace environment is essential to worker productivity and satisfaction. Agency leaders at all levels should develop strategies to foster S&T workforce diversity and inclusion, which can fuel S&T innovation. First-line managers are often aware of diversity and inclusion issues and challenges faced by employees, but lines of communication with higher-level management are not always accessible. By fostering communication and creating better channels for open dialogue, agency leadership can rapidly respond to feedback that enhances an inclusive environment and strengthens recruitment and retention of Federal S&T talent. Agencies should identify and share best practices to promote S&T workforce diversity and a culture of inclusion and S&T innovation. Some suggestions include S&T mentoring programs to broaden participation of underrepresented groups in the S&T workforce and encouraging participation in employee resource and affinity groups. Human capital advisors (chief human capital officer or top human resource leaders at agencies) are critical members of the leadership team to support employee engagement and enable workforce planning through optimum use of human capital flexibilities to meet changing S&T workforce needs.

IDEA Lab, An Innovation Office at the Department of Health and Human Services

---

[17] Inclusion is defined as a culture that connects each employee to the organization; encourages collaboration, flexibility, and fairness; and leverages diversity throughout the organization so that all individuals are able to participate and contribute to their full potential. Taken from 2011 Government-Wide Diversity and Inclusion Strategic Plan: //www.opm.gov/policy-data-oversight/diversity-and-inclusion/reports/governmentwidedistrategicplan.pdf.

[18] Yang, Yang, and Alison M. Konrad. 2011. "Diversity and organizational innovation: The role of employee involvement." *Journal of Organizational Behavior* no. 32 (8):1062-1083.

[19] Nelson, Beryl. 2014. "The data on diversity." *Communications of the ACM* no. 57 (11):86-95

The Department of Health and Human Services (HHS) established the IDEA Lab in 2013 to promote innovation within the agency and improve the agency's ability to deliver its mission. The IDEA Lab stemmed from HHS leadership recognizing the need for a concerted effort to institute an innovation culture and to tackle cross-cutting challenges across the organization. As a solution, the IDEA Lab supports the HHS workforce in developing strategies using a suite of technologies to drive innovative problem solving. The IDEA Lab cultivates innovation throughout the agency, including the identification, testing, and validation of new ideas. Among its goals are the promotion of internal entrepreneurship, bringing outside talent into Government, and tackling cross-department issues of priority, such as modernizing acquisition and making health data public. Through its various programs, the IDEA Lab helps to break down institutional siloes, in which innovative ideas can dissipate, and identify opportunities across the agency to implement new ideas.

Source: HHS, "IDEA Lab," http://www.hhs.gov/idealab.

The U.S. Global Development Lab at the United States Agency for International Development

Building on the belief that science, technology, innovation, and partnership can accelerate global development faster, cheaper, and more sustainably, the United States Agency for International Development (USAID) established the U.S. Global Development Lab (The Lab) in April 2014. The Lab has a two part mission:

- To produce breakthrough development innovations by sourcing, testing, and scaling proven solutions to reach hundreds of millions of people.
- To accelerate the transformation of the development enterprise by opening development to people everywhere with good ideas, promoting new and deepening existing partnerships, bringing data and evidence to bear, and harnessing scientific and technological advances.

A core function of the Lab is to devise, propose and test innovative ways to do business better, faster, cheaper, and more sustainably, and then move these innovations into practice across the Agency. Such innovations include the use of new, flexible hiring authorities and expanded fellowship opportunities to attract specialized talent not usual in government and the use of new and flexible procurement vehicles which allow for increased co-creation with our partners and non-traditional development actors.

Source: Department of State, United States Agency for International Development.

DOE STEM Mentoring Program

In 2009, President Obama created an executive order resulting in a Federal initiative to increase the participation of underrepresented groups, including women and girls, in the fields of Science, Technology, Engineering and Mathematics (STEM).

The Department of Energy's (DOE) STEM Mentoring Program was established in September 2010. To support this initiative, DOE employees are partnered with college students and encouraged to share their professional and academic experience in an effort to encourage underrepresented groups to become interested in and pursue careers in STEM fields.

The STEM Mentoring Program has established a partnership with area high schools to help identify and address student and facility needs relating to STEM.

Source: Department of Energy.

Leadership of agencies at all levels should identify and integrate best practices to support the flow of Federal S&T workers within and across agencies and sectors, including the private sector and academia, to facilitate recruitment, retain top talent, and foster innovation. Recruiting and retaining top S&T talent and creating a culture of S&T innovation depend on opportunities for the S&T workforce to advance technical skills and develop professionally through exposure to innovative practices and ideas. Federal leaders currently offer such opportunities by rotating S&T workers to other Federal agencies and the private sector, encouraging attendance at conferences and S&T meetings, and promoting a culture of continuous learning. Additionally, leaders should foster in-person collaborations through interagency agreements to exchange S&T talent across the Federal Government, as well as detail and exchange experiences across sectors.

Practice #1 – Enable Leaders to Create a Culture for Inclusion

S&T workers want an inclusive, diverse, and dynamic environment where their skills are valued and rewarded. The goals of an inclusive workplace climate include attracting new talent, maximizing employees' potential, supporting the workforce to further the mission of the agency, and improving employee retention. Harnessing talent from the diverse S&T workforce is essential to work that leads to creative discoveries and innovative solutions to the challenges faced by the Federal S&T community. Promoting inclusion requires savvy leadership to set the tone for the workplace environment and culture. Leaders need training in methods to create a culture of inclusion within their agencies.

The National Institutes of Health Office of Equity, Diversity, and Inclusion

Agencies are taking proactive steps to build a culture of inclusion. For example, the National Institutes of Health (NIH) Office of Equity, Diversity, and Inclusion has the mission to "cultivate a culture of inclusion where diverse talent is leveraged to advance health discovery." NIH commissioned the Office to design Special Emphasis Portfolios to highlight positive, equitable, and inclusive employment experiences for federally identified minorities. These portfolios are an integral part of the success in creating an inclusive environment at NIH.

Source: NIH, Office of Equity, Diversity, and Inclusion, http://edi.nih.gov/.

Practice #2 – Promote Career Flexibility and Mobility

Mobility of S&T talent across the Federal, academic, and private sectors helps to bring fresh ideas and perspectives to effectively solving the complex challenges facing the Nation. Barriers to attracting this talent may be reduced for the short term through one- or two-year details. Established programs such as the Department of Defense's Information Technology Exchange Program (ITEP) and the Department of Homeland Security's Loaned Executive Program (LEP) may serve as the foundation for other S&T areas of national importance. Agency leadership can facilitate interagency exchange programs to enhance mission critical skill development and optimize enterprise-wide skill utilization by detailing workers for 1 to 2-year rotations in another agency.

The Department of Homeland Security Loaned Executive Program

The Department of Homeland Security (DHS) Loaned Executive Program (LEP) was established to provide top executive-level and subject matter experts from the private sector an (unpaid) opportunity to share their expertise and provide innovative solutions to homeland security challenges with the Department. LEP participants are recruited from for-profit and nonprofit private-sector entities.

Source: http://www.dhs.gov/loaned-executive-program.

Intelligence Community Joint Duty Program

Authorized by Congress in the Intelligence Reform and Terrorism Prevention Act of 2004, the Intelligence Community (IC) Civilian Joint Duty Program is designed to help the IC, including its S&T workforce, with the development of cross-agency expertise through rotational assignments that foster a diverse and dynamic environment of information sharing, interagency cooperation, and intelligence integration.

Recent expansion of this program to include the GS-11 and GS-12 level may provide additional opportunities for the IC's S&T workforce to enhance their career development by embracing this successful human capital flexibility that is further shaping an integrated IC workforce.

Program goals and policies for joint duty were further outlined in 2013 by the Director of National Intelligence (ODNI), James R. Clapper, when he issued an updated Intelligence Community Policy

Source: ODNI, IC Joint Duty Policy and Guidance, http://icjointduty.gov/docs/ICD_660.pdf.

Information Technology Exchange Program

Information Technology Exchange Program (ITEP) is a mechanism for temporary exchange of information technology employees from the Department of Defense and the private sector. ITEP began as a pilot established by the National Defense Authorization Act (NDAA) for Fiscal Year 2010 and was extended to 2018 by the NDAA for Fiscal Year 2014. ITEP can support career mobility for current employees while serving to meet short-term workforce or mission needs. ITEP may provide needed career flexibility, attract valuable S&T talent from the private sector into Federal service, and facilitate collaborations and exchange of ideas across sectors. Effective implementation of the program has resulted in agency-generated guidance to assist in navigating the complex requirements of transitioning between the private sector and Federal service.

Source: NDAA of Fiscal Year 2010, Sec. 1110. *Pilot Program for the Temporary Exchange of Information Technology Personnel.*

S&T Workforce Engagement in and Alignment to Mission

Agencies should cultivate strong connections between the Federal S&T workforce and their missions.

When fostered by invested and focused leadership, an engaged workforce will yield innovative solutions to the world's greatest problems and will further exploration through world-class science. Employee engagement and motivation is generally associated with higher-quality products, increased output, and higher employee retention. Effective engagement of the Federal S&T workforce can be enhanced by properly identifying, acknowledging, and articulating the importance of its work. Leadership can contribute by being sensitive to S&T workers' unique career development needs, such as flexibility and mobility; fostering in-person collaboration, including attendance at professional conferences and maintaining affiliations with professional societies; and developing attractive workplace attributes that enable and motivate the S&T workforce to fully engage within their organizations at all career stages.

Because recognition of performance is a key component of engagement, the FSTW also recommends that the leaders of agencies at all levels review their incentive and reward systems to align them, to the extent possible, with both S&T workers' values and mission goals. For example, recognition of publications and patents relevant to mission accomplishment aligns well with professional values of S&T professionals. There may also be mission-unique products that do not fit as easily into the value system of S&T professionals that could be promoted in the broader S&T community, creating a different type of recognition. To improve recognition of the S&T workforce, the FSTW recommends that agencies explore modification of current incentive and reward systems for S&T workers to create new approaches that have the flexibility to recognize S&T professionals working autonomously, as well as incentivizing teamwork and innovation to solve mission-specific challenges. Agencies should coordinate and share best practices for programs that demonstrate effectiveness.

Professional development is another key component of engagement. Agency leaders should encourage employees to develop skills that enhance the agency's ability to accomplish missions in a dynamic and changing environment. One of the greatest challenges that S&T subject matter experts face is staying current with the technical expertise in dynamic fields of science and engineering. Those employees with advanced degrees in technical disciplines need encouragement to maintain their qualifications and broaden and enhance their skills in ways that optimally benefit their agencies, such as through support of their involvement in professional societies and participation in professional conferences and workshops.

Another aspect of aligning the S&T workforce with the agency mission is to ensure that the mission and most immediate goals of the agency are reflected in the structure of the workforce. Agency leaders should make S&T workforce planning a priority and allocate management time to connecting workforce planning efforts with strategic S&T program plans. Workforce planning reflects a long-term investment in the Federal workforce and provides a vision for the S&T workforce. S&T leaders must be at the center of workforce planning activities. Aligning workforce planning with specific S&T strategic plans is essential to maintaining the optimal and agile workforce needed as technology and science evolve with the changing needs of the Nation. In dynamic, innovative technology organizations, the S&T workforce is connected to the mission of the organization and the ability to meet stakeholder needs.

The FSTW recommends that agency human capital advisors and agency leaders work with the U.S. Office of Personnel Management (OPM) to support their organizations in providing S&T hiring managers with periodic analysis of and access to relevant data on the agency and government-wide S&T workforce. This practice can better inform S&T workforce managers of progress in meeting strategic workforce planning and engagement goals. Agencies must enable the development and use of metrics, measures, and tools to better characterize S&T employee engagement needs and to develop understanding of areas where there may be critical gaps in S&T skills. Specific data that could be useful for this purpose include data related to performance management, onboarding, retirement eligibility, hire and loss patterns, and grade and pay plan structures.

Workforce planning efforts should shed light on where skills gaps exist or may develop in an agency, and both leaders and employees should be empowered to fill those gaps through skills expansion. The FSTW recommends that agencies develop and, in partnership with OPM, promulgate innovative approaches to credentialing S&T skills that capture mastery rather than degree attainment to better align S&T workers with mission needs. Agencies should evaluate the qualifications of S&T applicants and guide professional development of the existent S&T workforce based on S&T skill.

Practice #1 – Align and Identify Critical Skills through Strategic Planning

Leaders within Federal agencies need to envision a dynamic, diverse, and high-caliber Federal S&T workforce whose multi-disciplinary skills will provide the professional, technical, and policy expertise to achieve their wide-ranging missions. Federal Government leaders should work to proactively drive efforts to engage the S&T workforce in the agency mission and identify the S&T workforce as a valuable contributor to that mission. This includes a workforce strategy that identifies needed skills (both scientific and related others) to meet agency missions and mechanisms to engage with and develop the current S&T workforce to meet those needs.

Workforce Strategy at National Aeronautics and Space Administration's Langley Research Center

Leaders at the National Aeronautics and Space Administration's (NASA) Langley Research Center supported S&T and human resource professionals to develop a comprehensive, multi-year workforce transformation plan that is strategically aligned with current and future mission priorities. The goal of this strategic workforce planning process was to optimize the Langley Research Center workforce in maximizing the performance and sustainability of the NASA mission. The approach taken was to maintain and provide the right skills at the right time, while also committing to invest in the workforce to build skills in emerging areas for future opportunities.

Source: Presentation by NASA Langley Strategic Workforce Planning Coordinator.

Centers for Disease Control and Prevention's Informatics Training in Place Program

To assist the Centers for Disease Control and Prevention (CDC) and State, Tribal, Local, Territorial (STLT) health departments (HDs) address gaps in public health informatics capacity and skills, CDC has developed the Informatics Training-In-Place Program (I-TIPP). I-TIPP is an innovative and cost effective approach for training CDC and STLT health department staff in public health informatics skills while they remain on their jobs. I-TIPP uses a virtual, on-the-job, applied training curriculum in the workplace. Launched in 2013, I-TIPP: (1) has provided opportunities for fellows to form a nationwide peer-to-peer network; (2) develops a cadre of informatics mentors in public health agencies; (3) serves as a platform for building informatics units and leadership in public health agencies; and (4) allows I-TIPP fellows placed in different settings (e.g., health care and public health) to collaborate across boundaries and achieve outcomes larger than what one fellow could achieve.

Source: Centers for Disease Control and Prevention.

The Environmental Protection Agency's Strategic Research Action Plans

> Beginning in 2011, the Environmental Protection Agency's (EPA) redesigned its research programs to enable the Office of Research and Development (ORD) to better respond to the Agency's priorities and to advance the science of sustainability. This effort transformed EPA's research portfolio form 13 research programs into six coordinated and highly integrated national research programs: Air, Climate and Energy; Chemical Safety for Sustainability; Homeland Security Research; Human Health Risk Assessment; Safe and Sustainable Water Resources; and Sustainable and Healthy Communities. In designing each of the six national research programs, EPA scientists ushered in a new era of research planning based on transparent, public engagement with partners and stakeholders across the Agency and scientific community. This new approach to research planning has created many opportunities to collaborate, leverage expertise, and coordinate research among areas that were previously planned and managed independently. The results of this planning process are outlined in a series of *Strategic Research Action Plans*, one for each of the six national research programs. These plans form the foundation and establish the time horizon for ORD's workforce planning process. As a result, ORD is in a position to make better informed decisions, such as what type of hiring mechanisms to use depending on the type of expertise needed and how long that expertise will be needed. For example, to respond to needs to obtain new talent in emerging science areas for a short period of time, ORD can utilize mechanisms such as the Federal Postdoctoral Hiring Authority.
>
> Source: Environmental Protection Agency.

Practice #2 – Analyze Engagement Data and Assess Feedback

Agencies have efforts underway to use data analytics to better understand the drivers and level of engagement with the agency mission within the S&T Federal workforce. The President's Management Agenda cross-agency priority goal in the area of people and culture and the joint-Executive Memorandum *Strengthening Employee Engagement and Organizational Performance*, both of which aim to foster a culture of engagement and excellence in the Federal workforce to enable high performance, are driving some of these efforts.[20] The Federal Employee Viewpoint Survey (FEVS), an annual survey that measures employees' perceptions on various characteristics of their organizations such as leadership and job satisfaction, and other data collected by agencies provide a wealth of information to better analyze trends and the current status of engagement with the S&T workforce.[21] Random, web-based inquiries that pop up when employees are on the system network are useful ways to receive feedback on day-to-day employee engagement. Stay interviews, whereby current employees are interviewed about their reasons for remaining with an agency, are another mechanism for collecting feedback on the impact of new agency policies, such as travel and conference regulations, on employee performance and morale. Stay interviews with high-potential and top S&T performers provide information on engagement efforts that are working well for employees. Acts of recognition such as agency awards and medals reward employee efforts to meet an agency's mission needs, and provide a mechanism to incentivize the S&T workforce to strive for excellence.

---

[20] Performance.gov, "People and Culture," available at http://www.performance.gov/content/people-and-culture#overview; Office of Personnel Management Memorandum, Strengthening Employee Engagement and Organizational Performance, Memo M-15-04, December 23, 2014.

[21] Office of Personnel Management (OPM), *Millennials: Finding Opportunity in Federal Service,* Federal Employee Viewpoint Survey Results. 2014.

Develop and Use Data Tools to Foster a Culture of Engagement

In July 2014, Office of Personnel Management (OPM), in partnership with Federal agencies, launched the Unlocking Federal Talent Dashboard (UnlockTalent) as part of the FY 2015 President's Management Agenda for "attracting, developing, and retaining the best talent in the Federal workforce." This new data tool is interactive and aims to enable agencies to better understand the valuable information provided through the Federal Employee Viewpoint Survey and other human resources data.

UnlockTalent, although still a fairly new initiative, provides the ability to analyze workforce data related to engagement and culture across the Federal Government. It can help identify successes and failures throughout an organization and allows agencies to benchmark with other parts of the Federal Government. The dashboard supports the needs for Federal leaders to improve understanding of their S&T workforce and identify strategies to better align incentive and reward systems with S&T employees' values.

Source: OPM, "UnlockTalent," https://www.unlocktalent.gov.

Employee Stay Interviews

The Department of Transportation (DOT), the Department of Homeland Security (DHS), and the United States Patent and Trademark Office (USPTO) are some of the agencies that have utilized stay interviews as one of their strategies for tactical things leaders could do to improve engagement. Stay interviews are structured, one-on-one conversations between supervisors and their direct reports designed to collect actionable data to strengthen employees' engagement and retention with the organization. They also help to identify and minimize any "triggers" that might cause them to consider leaving the organization. Stay interviews are different from exit interviews in that they empower supervisors to have conversations about what their employees enjoy about their work as well as identify barriers to mission accomplishment and suggestions for addressing those barriers. Once those factors are identified, supervisors can then reinforce or amplify the motivating factors and attempt to remedy issues before they become pain points for employees and thus help improve retention.

Federal agency leaders and human resources practitioners can utilize the valuable data collected from stay interviews to better understand their workforce and support data-driven decision-making regarding recruiting and retention activities.

Source: Department of Homeland Security.

Department of Commerce Honor Awards

Since 1949, the Department of Commerce (DOC) has granted honor awards in the form of Gold, Silver, and Bronze Medals. The Gold and Silver Medals are the highest and second-highest honor granted by the Secretary of Commerce for distinguished and exceptional performance, respectively. The Bronze Medal is granted by the head of an operating unit or Secretarial Officer for superior performance.

To warrant a Medal, a contribution must demonstrate performance at the highest levels in the execution, achievement or advancement of the mission of the Department of Commerce. Medals are awarded in a variety of categories ranging from scientific/engineering achievement to leadership to customer service to heroism. Medal winners receive a framed certificate and are celebrated at an annual awards ceremony.

Source: Department of Commerce.

Effective Authorities to Enable a Flexible and Agile Workforce

Agencies should make use of successful human capital flexibilities to shape the Federal S&T workforce.

An agile S&T workforce is needed to effectively respond to emerging R&D opportunities while enabling cutting-edge R&D to be performed and applied. Thus, creating an agile workforce system is essential for the Federal Government to remain competitive with other sectors for top S&T talent and reduce barriers to mobility of S&T talent to meet short- and long-term S&T mission needs. Working with OPM, agencies can expand their ability to use existing authorities optimally to create an effective and agile workforce system. While many authorities exist across the Federal Government or specific to an agency, there are constraints to their use and numerous authorities are underused. It is important for the Federal Government to continue the strategic use of existing authorities while continuously assessing how current flexibilities could be revised or how new flexibilities could be created to help improve Federal recruitment, hiring, and retention.

However, using existing authorities is insufficient to the challenge of creating an agile workforce system. Therefore, the FSTW recommends that the Director of OPM, in partnership with S&T agencies, identify proposals for new or expanded authorities that may help address specific human capital challenges and enhance agency competitiveness in hiring and retaining a world-class S&T workforce. Proposals should include the effective use and expansion of successful compensation flexibilities and full use of Federal cross-sector exchange programs.

Although the size of the Federal S&T workforce has increased since 2000, agencies face increasing retirements. Additionally, expectations from more recent generations of employees reflect the preference for a mobile career track that includes multiple employers and working in dynamic environments across sectors. Additional benefits and flexibilities common in non-Federal sectors, such as those that promote work/life balance and competitive compensation, effectively attract top S&T talent and affect the ability of the Federal Government to compete, which undermines agencies' abilities to shape the Federal S&T workforce to best meet Federal S&T missions.

Practice #1 – Offer Competitive Pay

Compensation is an important factor in career decisions. Although Federal service offers a multitude of benefits to attract S&T talent, including the public service mission and the importance of working in areas of national priority, compensation remains an important consideration in accepting an offer or staying in an organization. Federal agencies have substantial discretionary authority to provide additional direct compensation in certain circumstances to support their recruitment and retention efforts for the S&T workforce.[22] For example, the National Security Agency (NSA) increased salaries for most of its STEM positions by an average of 12 percent in September 2014 in an effort to recruit top STEM talent.

The Critical Position Pay Authority (CPPA) is a Federal Government-wide authority to increase base pay for critical positions. OSTP, OPM, and others identified the CPPA as a potentially underutilized flexibility, and in October 2014, OPM issued guidance to Chief Human Capital Officers on the flexibility.[23] Raising

---

[22] OPM, "Pay and Salaries," available at www.opm.gov/policy-data-oversight/pay-leave/pay-and-leave-flexibilities-for-recruitment-and-retention.

[23] Archuleta, K. "Memorandum for Chief Human Capital Officers: Critical Position Pay Authority," October 8, 2014; Peña and Mineiro, *History of the Critical Position Pay Authority*.

awareness of when the use of CPPA and other Federal compensation authorities is appropriate would help the Federal Government achieve a world-class S&T workforce. Agencies should ensure that any compensation flexibilities are used judiciously and in accordance with applicable law, regulations, agency policy, and budgetary limitations to enhance the Federal Government's ability to attract and retain top S&T talent.

Critical Position Pay Authority and Agency-Specific Variants

In 1990, Congress passed the Critical Position Pay Authority (CPPA) as a Federal Government-wide authority to facilitate recruitment and retention and address the pay disparities between the Federal Government and other sectors. The CPPA authorizes agencies to fix the base pay of select positions identified as "critical" in a scientific, technical, professional, or administrative field at a rate higher than would be otherwise payable. The CPPA is approved for 800 Federal positions, yet the CPPA has only been used for 34 positions and the majority (24 or 71 percent) were positions in the Federal Bureau of Investigation (FBI). The CPPA also serves as a model for other agency-specific authorities, such as NASA's Critical Position Pay Authority.

The effective use of the CPPA depends on various factors, including the Federal hiring community's awareness of the authority and careful navigation of complex regulations and justifications for its use, but efforts in these areas may be lacking. Ongoing barriers to the authority's use may be mitigated through rigorous review of its effectiveness, pilots to test potentially beneficial regulatory revisions, and interagency efforts to provide guidance.

Source: 5 U.S.C. § 5377; 5 CFR § 535; V. Peña and M. C. Mineiro, *History of the Critical Position Pay Authority and Options to Support Its Use* (Alexandria, VA: Institute for Defense Analyses, march 2014), IDA Document D-5159.

Practice #2 – Convert Staff in Temporary Positions to Permanent Federal Service

Every year, talented S&T employees participate in a variety of Federal and non-Federal internship and fellowship opportunities that temporarily place them throughout the Federal Government. To increase flexibility within the Federal Government to use such programs as hiring tools, some Federal programs, like the Pathways Programs for students and recent graduates, are governed by a special hiring authority that allows for the non-competitive conversion of talent into Federal service.

Pathways Non-Competitive Conversion Authority

The Pathways Programs are developmental programs aimed at promoting opportunities for students and recent graduates into Federal service. The programs include the Internship Program, Recent Graduates Program, and the Presidential Management Fellows Program, which provide short-term appointments at agencies. The Pathways Programs is governed by a Federal Government-wide excepted service authority that allows interns to be hired (i.e., converted to regular employee status) on a noncompetitive basis.

Many agencies depend on this authority to enhance their ability to recruit promising students and recent graduates into their workforce and several non-Federal fellowship opportunities exist that also aim to foster Federal opportunities for students and recent graduates with training in S&T fields.

Source: Executive Order No. 13562, *Recruiting and Hiring Students and Recent Graduates*; 5 CFR § 213.3402(a),(b),(c), Schedule A.

Presidential Innovation Fellowship Program

Presidential Innovation Fellowship (PIF), like the ITEP, a Federal exchange program that targets bringing in S&T talent from the private sector for short terms of service, the PIF was established by the White House in 2012 and is now operated by the General Services Administration. The program brings innovators from the private sector for 12 months throughout various parts of the Federal Government. Fellows are hired as Federal employees for that time period and work on pressing challenges facing agencies.

Some examples of projects that the fellows have initiated include enabling online tools to make it easier to do business with the Federal Government, such as responding to Request for Proposals, and building apps to engage the public in crowdsourcing information to support disaster planning and preparedness. Hiring of fellows is governed by a Federal Government-wide excepted service authority and is on a non-competitive basis, which facilitates their entry, albeit temporary, into Federal service.

Source: The White House, "Executive Order – Presidential Innovation Fellows Program" available at https://www.whitehouse.gov/the-press-office/2015/08/17/executive-order-presidential-innovation-fellows-program.

Effective Relationships to Meet Strategic Workforce Needs

Agencies should improve collaborations between S&T professionals and their human capital counterparts (chief human capital officer or top human resource leaders at agencies) to strengthen human capital practices and enhance quality, diversity, and flow of the Federal S&T workforce.

The Federal Government's ability to continue to innovate and advance S&T achievement depends on the Nation's ability to attract and retain top S&T talent into the Federal workforce. To achieve this, in addition to engaging in other human capital practices like workforce planning and succession planning, agency leaders should work to strengthen collaborations between S&T professionals and human capital advisors. Agencies should seek opportunities to share best practices and work together to optimize recruitment and hiring practices, such as the use of market research to inform hiring strategies and the development of clear position descriptions to more effectively attract S&T talent and compete with the private sector. This includes sustaining support of a human capital community of practice that continues to strengthen, adopt, adapt, and learn from successes in human capital practice and have enhanced quality, diversity, and flow in the Federal S&T workforce.

Through a community of practice, human capital advisors can create and issue shared guidance on use of phased retirement or emeritus programs to keep S&T retirees engaged in the workforce and mentoring younger S&T workers, which is important to ensuring transfer of critical scientific and technical information. Human capital advisors can also share and exchange best practices for conducting stay interviews and exit interviews to inform workforce planning and address S&T workforce concerns. In addition, the community of practice can support agency efforts to share résumés of S&T candidates, including those participating in Pathways Programs and others eligible for non-competitive conversion. A community of practice is also useful to keep all agencies abreast of key issues involving potential skills gaps in the S&T workforce. Employing these various practices within agencies will help lay a foundation for strengthening the Federal S&T workforce.

In addition to advice from in-house human capital advisors and OPM, agencies should consider seeking advice from external S&T committees, councils, and societies on opportunities to improve human capital practices. Additionally, private sector efforts to characterize skills gaps, broader workforce attributes, and employment matching can be valuable to the government sector.

Practice #1: Seek External Advice

To better understand practices used by the business and academic communities, agencies have used external advisory committees to seek unbiased, sound, and balanced advice regarding human capital practices such as workforce planning and succession planning. This approach can offer an agency an industrial and academic perspective that has proven valuable to those agencies which have engaged an advisory committee.

Seeking External Advice on Workforce Planning

The National Science Foundation (NSF) established a committee to advise the Foundation on business and operations management, including issues within human capital practice under the Federal Advisory Committee Act (FACA). Committee members are from outside the Federal Government in the areas of scientific research administration, human resources management, information technology, government performance, education management, academia, and business. They report to NSF not only on business practices, but also on human capital practices.

Source: www.nsf.gov/oirm/bocomm.

Visiting Committee on Advanced Technology

The National Institute of Standards and Technology (NIST) is advised by its Visiting Committee on Advanced Technology (VCAT), an advisory body in accordance with the FACA. The VCAT reviews and makes recommendations regarding general policy for NIST, its organization, its budget, and its programs, including strategic human capital management. The VCAT comprises 15 members, appointed by the NIST Director, from industry and academia who serve three year terms. In its official role as the private sector policy advisor of the Institute, the VCAT's annual report identifies areas of research and research techniques of potential importance to the long-term competitiveness of United States industry, and areas in which NIST possesses special competence, which could be used to assist United States enterprises and United States industrial joint research and development ventures.

Source: Department of Commerce, National Institute of Standards and Technology.

Appendix: Full List of Recommendations

FSTW members emphasized several areas of importance for ongoing discussion and consideration by a more permanent interagency effort. This appendix provides a full list of recommendations and actions discussed by the FSTW members organized by overarching priority, including those mentioned in this report.

1. **Empower the capacity of the Federal S&T workforce to sustain and promote a world-class science, technology, and engineering enterprise through thoughtful, proactive leadership.**

    1.1. Leadership of agencies at all levels (senior executives, S&T managers, etc.) should make S&T workforce planning a priority and devote management time to recognizing that workforce planning is a long-term investment. Workforce planning efforts should be directly linked to implementing the agency's strategic S&T plans.

    1.2. Leadership of departments and agencies at all levels should enable, encourage, and engage in workforce planning efforts. Workforce planning performed by agency leaders should include S&T leadership to forecast the changing landscape of S&T fields, skills, and workforce capabilities, and should be informed by human capital advisors (chief human capital officer or top human resource leaders at departments and agencies) to provide guidance to those that manage workforce planning on how to optimally utilize existing human capital flexibilities to meet changing S&T workforce needs.

    1.3. Leadership of agencies at all levels should identify and integrate best practices to foster flow of Federal S&T workers across the Federal government, such as facilitating recruitment and retention of S&T talent by developing their technical skills to maintain the mission-critical S&T capabilities for the Federal S&T enterprise.

    1.4. Leadership of agencies at all levels should develop strategies to increase S&T workforce diversity, inclusion, and a culture of S&T innovation within their workplace environments. Some challenges faced by the S&T workforce include recruiting from a less-diverse pool of qualified applicants, need for in-person collaboration, and managing opportunities for flow to optimize exposure to innovative practices and talent. Leadership at all levels should work to implement these strategies.

        1.4.1   Agencies should identify and share best practices to promote S&T workforce diversity and a culture of S&T innovation, including S&T mentoring programs to broaden participation of underrepresented groups in the S&T workforce and encouraging participation in employee resource and affinity groups.

        1.4.2   The Director of OPM, working with the agencies, should develop leadership training and guidance on ways to embrace workforce diversity and inclusion that meets the needs of the S&T workforce and enhances a culture of S&T innovation. In addition, the Director of OPM should work with the Equal Employment Opportunity Commission to develop best practices for training to reduce unconscious bias facing S&T applicants and existing S&T workers.

1.4.3 The Director of OPM, working with agencies, should revise leadership programs and leaders' performance evaluation criteria to assess the inclusiveness of their work environments. Leaders should have incentivizes to recommend opportunities to improve upon the workplace culture as part of their performance evaluation criteria.

1.5. Leadership of agencies at all levels should review current incentive and rewards systems to understand how they are aligned with S&T workers' values (e.g., rewards for publications or patenting) and should revise current systems for S&T workers to establish rewards based on S&T workers' values, as appropriate.

1.6. Leadership of agencies at all levels should identify and integrate best practices to support the flow of Federal S&T workers across sectors, including the private sector and academia, to facilitate recruitment, retain top talent, and foster innovation. Recruiting and retaining top S&T talent and creating a culture of S&T innovation depend on prioritizing opportunities for the S&T workforce to advance technical skills and professional development (e.g. conferences and S&T meetings), promote a culture of continuous learning, and foster in-person collaborations through interagency agreements to exchange S&T talent across the Federal Government as well as detail and exchange experiences across sectors.

1.7. Leadership of agencies at all levels should expand their use of active recruiting to identify, target and bring aboard the best candidates, especially for hard-to-fill S&T positions. Leading organizations of all kinds deploy active recruiting based on the belief that exceptionally talented individuals are typically already employed and not actively looking, but nonetheless can be persuaded to leave for the right role. All agencies currently have the authority to deploy active recruiting best practices for all jobs, including competitive service roles. For example, USAID's Global Development Lab currently uses these practices, which involve close collaboration between the recruiter and hiring manager to develop ideal candidate characteristics, research to develop candidate pools through interviewing subject matter experts and mining social media, outreach to promising candidates, and building and maintain outside talent networks that can sustain a pipeline of exceptional talent into an agency for a range of critical roles.

2. **Cultivate engagement in the Federal S&T workforce.**

    2.1. Agency leaders should identify, acknowledge, and articulate the role that their S&T workforce plays in meeting the department and agency missions.

    2.2. The Director of OPM, in partnership with agencies, should research systematic and sustainable practices on the use of analytics, data, metrics, and tools (e.g. tool kits, information resources, and clearinghouses) to support workforce forecasting in mission-critical competencies.

        2.2.1 Agency human capital advisors should work with OPM to support their institutions in providing S&T hiring managers with periodic analysis and access to relevant Federal Government-wide employee data (performance management, onboarding, retirement eligibility, grade structure, agency and Federal statistics, etc.) to inform S&T workforce management.

        2.2.2 Agencies, in partnership with their human capital advisors and OPM, should identify best practices across all sectors (including the Federal Government) in S&T workforce forecasting, including the use of metrics and measures to better align and predict S&T workforce needs with critical S&T capabilities, and how they can be applied within the Federal Government.

2.3. Agencies, in partnership with their S&T leaders, should seek external S&T workforce advice through the creation of new committees or the expansion of existing advisory groups (e.g. Federal advisory committees).

2.4. Agencies, in partnership with their human capital advisors, should identify ways to revise the occupation classification system to capture sub-disciplines to more accurately assess S&T workforce gaps and provide recommendations to the Director of OPM on how to improve the classifications system to improve S&T workforce capacity.

2.5. Agencies should develop and, in partnership with OPM, promulgate innovative approaches to credentialing S&T skills that capture mastery rather than degree attainment to better align S&T workers with mission needs. Departments and agencies should evaluate the qualifications of S&T applicants and guide professional development of the existent S&T workforce based on S&T skill.

2.6. Agencies, in partnership with their human capital advisors, should communicate the role of Federal human resource staff as strategic partners in workforce planning issues, including developing formal or informal training for human resource staff on S&T workforce planning, and enabling a more effective partnership between, and reciprocal understanding of S&T workforce needs by, human resource professionals and individuals in S&T positions, as appropriate.

3. **Use successful human capital flexibilities to shape the Federal S&T workforce.**

    3.1. The Director of OPM should continue to publish guidance to promote and assist in the full use of existing human capital flexibilities to enhance competitiveness, including technical assistance and support in developing evidence-based rationales that may be needed to authorize use of such flexibilities (e.g. direct hire, critical pay, special rates, student loan repayments, use of Schedule A authorities, non-permanent hiring of S&T talent, and recruitment, relocation, and retention incentives). For example, all agencies currently have the authority to use Schedule A (r) to hire technical talent for fellowships of up to four years using streamlined, flexible and accelerated hiring processes (e.g., USAID's Global Development Lab and GSA's 18F have both used this authority to build technical teams).

    3.2. The Director of OPM in partnership with S&T agency representatives on the NSTC should identify and explore proposals for new or expanded authorities that may help address specific human capital challenges and enhance agency competitiveness in hiring and retaining a world-class S&T workforce (e.g., exchange of S&T talent across sectors, compensation flexibilities, and expansion of the Pathways Programs' non-competitive conversion authority to non-Federal internship and fellowship programs).

    3.3. The Director of OPM, in partnership with agencies, should: (1) develop a clearinghouse of available human capital authorities, including examples of their use across the Federal Government and benefits and impacts to the S&T workforce, and (2) identify strategies and promulgate existing OPM guidance on implementing hiring flexibilities, such as superior qualifications and special needs pay settings, bonuses, and creditable service for leave accrual, to facilitate the hiring of and the maintenance of S&T talent into Federal service.

4. **Improve human capital practice to strengthen Federal S&T workforce quality.**

    4.1. Agencies should work with their human resource advisors to identify approaches to hire qualified non-U.S. citizens into S&T positions when no qualified U.S. citizen has applied.

    4.2. Agencies should work with their human capital advisors to exchange best practices for conducting stay and exit interviews or surveys and the subsequent analysis to inform workforce planning and management decisions.

    4.3. Agencies, in partnership with OPM and their human capital advisors, should develop and share effective succession management practices (e.g. phased retirements or voluntary separation to add flexibility in shaping the S&T workforce and transferring knowledge, or emeritus programs to keep S&T retirees engaged in the workforce). The Director of OPM, in partnership with departments and agencies, should develop guidance on effective succession management practices, including case studies and lessons learned from various sectors to maintain a world-class Federal S&T workforce.

    4.4. Agencies should identify critical S&T workforce skills gaps and work with the Director of OPM to build partnerships with educational institutions (e.g. community colleges or credentialing bodies) to develop training programs that address the gaps.

    4.5. The Director of OPM, in partnership with agencies, should identify effective practices to increase early (i.e. prior to post-secondary) awareness of S&T opportunities in the Federal Government, including partnerships with local communities, school districts, post-secondary institutions, and professional societies and in-person testaments that express the value of a Federal Government career experience.

    4.6. The NSTC, human capital advisors, and OPM should develop tools to share résumés of S&T candidates across agencies, including S&T candidates participating in the Pathways Programs and other S&T candidates eligible for non-competitive conversion.

    4.7. The Director of OPM in partnership the NSTC should facilitate ways to identify and share effective recruitment and hiring practices across sectors (e.g. the use of market research to inform hiring strategies and the development of successful position descriptions).

5. **Create a cross-cutting interagency community of practice to share innovations in Federal S&T workforce practices and to track progress of Federal S&T workforce efforts.**

    5.1. An interagency community of practice should conduct a periodic, formal Federal Government-wide S&T workforce planning analysis through the NSTC to support Federal- and agency-specific long-term workforce forecasting and planning needs.

    5.2. An interagency community should develop an S&T messaging approach to attract and recruit S&T talent across the Federal S&T enterprise. The Director of OPM should support the promulgation of a Federal Government-wide S&T mission brand across the S&T enterprise.

5.3. An interagency working group should develop training for human resource and hiring managers on how specific skills possessed by S&T trained workers can be applicable to a broad range of jobs, and understand how these skills are transferrable across positions.

5.4. An interagency community of practice should work to improve the collection of Federal S&T employee data to better assess current workforce skills.

5.5. An interagency community of practice should identify and implement effective mechanisms that improve the management of clearances for the S&T workforce to facilitate the mobility of S&T talent across agencies.

References

Carnevale, A. P. *The Workplace Realities*. Alexandria, VA: American Association of School Administrators (AASA). 2008. http://www.aasa.org/SchoolAdministratorArticle.aspx?id=6000.

Carnevale, A. P., N. Smith, and M. Melton. *STEM: Science, Technology, Engineering, Mathematics*. 2011. https://cew.georgetown.edu/report/stem/

Carnevale, A. P., N. Smith, and J. Strohl. *Help Wanted: Projections of Jobs and Education Requirements through 2018*. Washington, D.C.: Center on Education and the Workforce, Georgetown University. June 15, 2010. https://cew.georgetown.edu/report/help-wanted/.

Davidson, Paul. "Science, Tech Jobs Pay More, Lead in Growth." *USA Today* (July 13, 2011). http://usatoday30.usatoday.com/MONEY/usaedition/2011-07-14-Tech-jobs_ST_U.htm.

Department of Defense Science, Technology, Engineering & Mathematics: STEM Education & Outreach Strategic Plan, 2010–2014. http://www.ngcproject.org/sites/default/files/DoD-wide%20STEM%20Education%20and%20Outreach%20Strategic%20Plan%20-%202010%20-%202014.pdf. 2009.

Langdon, D., G. McKittrick, D. Beede, B. Khan, and M. Doms. *STEM: Good Jobs Now and for the Future*. ESA Issue Brief 03-11. July 2011. http://www.esa.doc.gov/sites/default/files/stemfinalyjuly14_1.pdf. Academy of Engineering and National Research Council of the National Academies. *An Interim Report on Assuring DOD a Strong Science, Technology, Engineering, and Mathematics (STEM) Workforce*. Washington, D.C.: National Academies Press. 2012. http://www.nap.edu/catalog/13433/an-interim-report-on-assuring-dod-a-strong-science-technology-engineering-and-mathematics-stem-workforce.

Office of Personnel Management (OPM). *Millennials: Finding Opportunity in Federal Service*. Federal Employee Viewpoint Survey Results. 2014.

OPM.gov. "2014 Federal Employee Viewpoint Survey," available at http://www.fedview.opm.gov.

Partnership and Universum USA, "Great Expectations! What Students Want in an Employer and How Federal Agencies Can Deliver It." 2010. http://ourpublicservice.org/publications/viewcontentdetails.php?id=129.

Peña, V. and M. C. Mineiro. *History of the Critical Position Pay Authority and Options to Support Its Use*. Alexandria, VA: Institute for Defense Analyses. IDA Document D-5159. March 2014.

Performance.gov, "People and Culture," available at http://www.performance.gov/content/people-and-culture#overview

Office of Personnel Management (OPM) Memorandum. "Strengthening Employee Engagement and Organizational Performance." M-15-04 (December 23, 2014).

Abbreviations

AAAS	American Association for the Advancement of Science
CDC	Centers Disease Control and Prevention
CPPA	Critical Position Pay Authority
DHS	Department of Homeland Security
DOC	Department of Commerce
DOE	Department of Energy
DOT	Department of Transportation
EPA	Environmental Protection Agency
FACA	Federal Advisory Committee Act
FBI	Federal Bureau of Investigation
FEVS	Federal Employee Viewpoint Survey
FSTW	Fast-Track Action Committee on the Federal S&T Workforce
HHS	Department of Health and Human Services
IC	Intelligence Community
ITEP	Information Technology Exchange Program
I-TIPP	Informatics Training-In-Place Program
LEP	Loaned Executive Program
NDAA	National Defense Authorization Act
NIH	National Institutes of Health
NIST	National Institute of Standards and Technology
NSA	National Security Agency
NSF	National Science Foundation
NSTC	National Science and Technology Council
ODNI	Office of the Director of National Intelligence
OPM	Office of Personnel Management
OSTP	Office of Science and Technology Policy
PIF	Presidential Innovation Fellowship Program
R&D	research and development
S&T	science and technology
STEM	science, technology, engineering, and mathematics
STLT	State, Tribal, Local, Territorial
USAID	United States Agency for International Development
USPTO	United States Patent and Trademark Office
VCAT	Visiting Committee on Advanced Technology

www.ingramcontent.com/pod-product-compliance
Lightning Source LLC
Chambersburg PA
CBHW080528190526
45169CB00008B/3094